Copyright © Kally Bougadis 2011

This is a work of fiction. Names, characters, places and incidents either are products of the author's imagination or are used fictitiously and any resemblance to actual events or locales or persons, living or dead is entirely coincidental.

All rights reserved. No part of this book may be used or reproduced in any manner whatsoever without written permission of the author or Speed Math Press except in the case of brief quotations embodied in critical articles and reviews.

**CalcaLand: The ODD Twins get EVEN!**
**ISBN-13: 978-0-984-09330-4**

Interior Layout: Custom-book-tique
*For bulk order prices or any other inquiries, please contact*
*customercare@speedmathpress.com*

**www.speedmathpress.com**

# CalcaLand

## The ODD Twins get EVEN!

Written by a Special Education teacher
Kally Bougadis

Illustrated by
Lilian Barac & Milos Trando

# Endorsements

Some students always struggled to learn the difference between odd and even numbers, CalcaLand reaches those struggling students.

—**Peter Tang, Elementary/Middle School Math teacher.**

I have a learning disability called ADHD, If I was taught math using CalcaLand in this fun creative way, I think I would have been declassified.

— **Rob Tamboia – Motivational Speaker, Author, Founder of Champion Life**

Kally Bougadis' book CalcaLand is a small gem filled with great insights.

—**Ruben Gonzalez Olympian, Speaker, Author of "The Courage to Succeed"**

Key to my success in becoming a Tenth Degree Black belt Champion was mastering the Fundamentals of Karate. CalcaLand teaches children how to master the fundamental skills that will help all students succeed in Math- A must Read!

—**Dr. Stan Harris – 10th Degree Black Belt, and Motivational Hall of Famer**

# Acknowledgements

To my wonderful parents and role models, James and Christine Bougadis, for giving me firm wings to soar. To my father for always making me feel that I could do no wrong and sparking my entrepreneurial spirit. To my mother for your insight, strength and instilling the value of education. I am eternally grateful for your unconditional love and support.

My love and thanks also to my sisters, Penny Tsekouras, Kathryn Bougadis and Soula Tsekouras, YOU are my everlasting cheerleaders. To my brother in law Pete Tsekouras, who pushed me to complete the book already! You mean the world to me, what would I ever do without you guys?

To my nephew Peter Tsekouras, my Godson Jimmy and my niece Christiana, whom I love so much and who represent the characters in my book.

To my cousin Kalli Papantoniou, Jimmy Liapis and Melissa Echiveri, you guys are my number one fans. To my friend Stacy Pino, thanks for connecting my words when I had writer's block.

To my dear angel Stephen Pierce, thank you for redirecting my steps to follow my heart, your consistent encouragement, inspiration and marketing expertise. I am eternally grateful to you.

To my princess Kai Williams, who inspired the book during a math lesson, and to her outstanding and continual improvement in math, which continues to make me proud.

To my student twins, who inspired the creation of Todd and Steven.

And to my extended D75 family and PS224, who have always honored, valued and inspired my teaching methods.

My phenomenal illustrators, Lilian Barac, Milos Trando, who colored my words in a way no one else could and my God-sent designer Maggie Pagratis.

My exceptional editors, Cheryl Stewart and Susan Brahinsky, who turned my script into a flowing waterfall.

To Rob Tamboia, Peter Tang, Ruben Gonzalez & Dr. Stan Harris for their inspiration and fantastic book reviews. Through you, I know that anything is possible.

To my Landmark and Tony Robbins community who empower and provide me with the tools to fulfill my dreams.

To the beautiful beaches of Croatia and Greece, where my script came to life.

To my student angels, you grant me purpose and joy. Who blossom despite the odds. Keep persevering and continue to make me proud.

And most importantly, to YOU, dear reader, and everyone with whom you choose to share this book with. Thank you to all my readers for contributing to our mission of empowering children with special needs.

And lastly, to my Lord, my light, my strength, my rock, who continues to guide me throughout my journey. Continue to lead and I will follow.

## Note to Reader

I deliberately designed bonus activities which model standardized state test questions and reinforce learning.

To receive bonus activities simply visit www.speedmathpress.com

### Recommendations

I highly recommend reading this book in two sittings. I would begin reading single digit numbers complete the activity pages and then move on to double digits to complete those activity pages.

### Activity Pages Include:

- Math Comprehension
- Sample Standardized State Testing Math Questions
- Math worksheets
- Reading Comprehension:
- Main idea
- Details
- Sequencing
- Drawing Inferences
- Comparing and Contrasting

Simply visit www.speedmathpress.com to receive these bonuses.

# Dedication

**To ALL my student angels.
I am so proud of each and everyone of you.**

On Calca Land,
Twins Todd and Steven always fight.
That's because each twin thinks he's right.

While Steven is the oldest by a few,
Todd always tells him what to do.
Steven doesn't like that
So, the boys argue.

Sarcastic, silly, small Steven
Simply said, prefers the even.
And he loves to divide and pair by two,
Without ever leaving out a few.
He conquers and can't stand remainders,
For Steven does not like these odd strangers.

Steven also feels that three's a crowd,
And since he doesn't want anyone to feel odd or blue,
He pairs his friends into even groups of two,
Then everyone always has something to do.

Now Todd doesn't mind being the odd one out,
So he chooses groups of 3, 5, 7, and 9,
And if someone feels alone and decides to pout,
Because they are the odd one out,
Todd will let them take his place.
Just so he can have some space.

Yes, some may say that Todd is rather odd.
But by odd, I don't mean strange, silly, or weird.
For odd is not a word that needs to be feared.
Todd is simply not like his twin brother,
He likes being odd and different from any other.

Odd just means the opposite of even.
When you count 1, 2, 3, you can easily see,
The numbers all line up as neat as can be.


So if 0 is even, then 1 must be odd,
And 2 must be even, which means 3 is odd.


Every other number is either even or odd,
And this greatly pleases both Steven and Todd.

So when the digits come knocking
on the mystical door,
They run to the twins that they adore.

0 laughs with Steven,
1 jogs with Todd.

2 sings with Steven,
3 plays with Todd.

Now if digits 5, 6, 7 and 8
Decide to show up a little late,

5 gallops with Todd,
And 6 pounces on Steven.

7 spins with Todd.
And 8 twirls with Steven.

Because 0, 2, 4, 6, and 8 are EVEN
and they play with Steven
While 1, 3, 5, 7 and 9 are Odd
and always play with Todd.

You see a number is always even or odd
And plays with either Steven or Todd.

Since 8 is even,

It will dance with Steven,

So why must 9 dine with Todd?

That's because

digit 9

is odd.

And since he is odd,

He will pair up with Todd.

So digits 1, 3, 5, 7 and 9,
Are numbers you can now define,
Because these digits are all odd,
They all line up to play with Todd.

While digits 0, 2, 4, 6, and 8,
Are numbers that are even.
They are easy to remember,
Because even rhymes with Steven.

There's also a leader who stands alone.
Upon a tall and golden throne,

The supreme ruler of Calca Land,
Is the one I hold secretly in my hand.
He's neither odd nor even,
And he doesn't play with Todd or Steven.

He is the leader of this numbers land,
And all numbers obey his every command.
King Calca is his honorable name.
And he's the ruler of this mathematical game.

Whenever the King enters Calca Land,
Everyone commemorates,
Bows down and celebrates
Because he's the one who magistrates.

King Calca decides who gets to play with whom,
And where each stands in this magical Calca Numbers Land.

One mysterious day,
The multi-digit number 2,736 came out to play.
And the twins had not one word to say!

This number had digits both even and odd,
So no one knew if it should play with Steven or Todd.

# Rules

Then along came King Calca on that fine day,
To help figure out the final say,
The King declared a rule that day.

And stated that in order for 2,736 to play,
It would be the end digit, the tag-a-long,
That would decide in which math group it would belong!

The King said the last digit is the one that counts,
No matter how large the number amounts.
The tag-a-long digit at the end of the train
Is the one that will determine that number's name.

So since the multi-digit number is 2,736
Has a number 6 digit at the very end,
Despite its odd and even blend,
The number must be even,
So it must play with sarcastic, silly, small Steven.

Now Steven is the multi-digit's favorite friend,
Because number 6 tags along at the very end.

But what if the number 345
decides to arrive and jive?
Who do you think will get to dance?

Yes, Todd will get the chance to jive
Because the number ends in...5.

What if 3,246,523 decides to flee?

Wouldn't you agree,
That if the twins start their chase,
Todd will have to win the race?

Why do you think this must be?

It's because the number ends in ...3.

3246523

Because 0, 2, 4, 6, and 8 are EVEN
and they play with Steven
While 1, 3, 5, 7 and 9 are Odd
and always play with Todd.

There's another reason the twins always fight,
It's because Todd thinks it's just not right
That zero is considered even.
He argues in a dreadful tone,
The digit should be left alone.

It's King Calca who settles all the duels,
For he's the one who makes the rules.
And he declares although you cannot pair the air,
If you divide 0 by two, there is never an odd remainder.
And since there is no leftover stranger, 0 is always even.
Therefore it gets to play with Steven.

When the twins begin to fight,
And can't stand to be in each other's sight,
They separate and walk apart.
They know taking a time-out is very smart.
Steven runs towards the even house numbers on the block,
And upon each door he chooses to knock.
Amongst the even numbers he chooses to flock.

Todd walks on the odd side of the avenue.
Doesn't it make sense that's where he'd run to?
He too, is in search of new friends on the block,
So upon each odd-numbered house, Todd will knock.

If Todd or Steven were walking on your street,
who would you likely meet?

King Calca created these basic rules
So the twins wouldn't have any future duels.
These are the simple laws of math.
Follow them closely and you will see
How easy Mathematics can be!

Now that you know the trick, I have no doubt,
That whatever number comes about,
You can figure it out,
Without getting worn out.

If you just remember...
In which group the final digit belongs,
And if it plays with Todd or Steven
And is considered odd or even.

Remember the math rules when grouping numbers,
And you will always get the answers right.
Then maybe, in time, the twins won't fight!

# About the Author

Kally Bougadis is an educator who mentors teachers and educates children classified with emotional disturbances. Her mission is to bridge the gap between special education and general education students through creativity and imagination, which is at the heart of every child regardless of classification. She uses the arts and fun techniques to engage students along with an easy dialogue that empowers them, transitioning their experience from one of struggle to striving in the learning process.

This book was inspired by one of her students, Kai Williams, who was struggling in math. She wrote the book to support *all* learners who are challenged in math and to teach fundamentals in a fun, easy way. Her commitment is that this book inspires educators to return to the roots of education--- educare, to "bring forth what is within."

Kally Bougadis holds a BA in Psychology from Queens College, a dual Masters in Special Education & Reading from St. John's University & a Building and School Leadership certificate from the College of St. Rose. She is a graduate of Landmark Education and serves as a School Board member of St. Nicholas Greek Orthodox Church.

She is a singer and songwriter of *Lead and I will Follow* and the founder of *SchoolCare Foundation*, Inc. which provides resources and support to manage the emotional well-being and educational growth of Special Education children with emotional needs. For more information, you can visit their website at www.schoolcarefoundation.org and www.speedmathpress.com

www.ingramcontent.com/pod-product-compliance
Lightning Source LLC
LaVergne TN
LVHW071027070426
835507LV00002B/49